Collins

INTERNATIONAL
PRIMARY
MATHS

Workbook 1

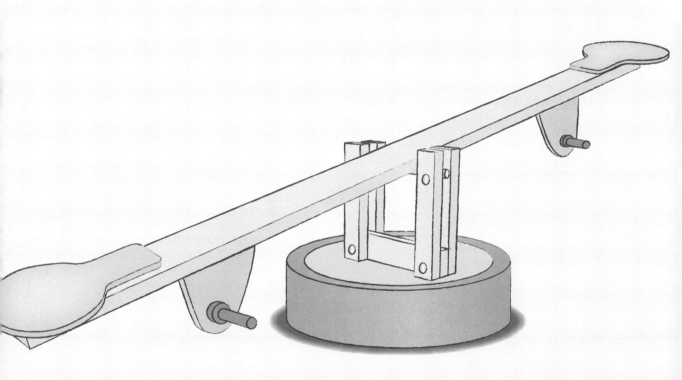

William Collins' dream of knowledge for all began with the publication of his first book in 1819. A self-educated mill worker, he not only enriched millions of lives, but also founded a flourishing publishing house. Today, staying true to this spirit, Collins books are packed with inspiration, innovation and practical expertise. They place you at the centre of a world of possibility and give you exactly what you need to explore it.

Collins. Freedom to teach.

Published by Collins
An imprint of HarperCollins*Publishers*
The News Building
1 London Bridge Street
London
SE1 9GF

HarperCollins*Publishers*
Macken House, 39/40, Mayor Street Upper,
Dublin, D01 C9W8,
Ireland

Browse the complete Collins catalogue at
www.collins.co.uk

10 9 8 7 6 5

ISBN 978-0-00-836945-3

British Library Cataloguing-in-Publication Data
A catalogue record for this publication is available from the British Library.

Author: Lisa Jarmin
Series editor: Peter Clarke
Publisher: Elaine Higgleton
Product developer: Holly Woolnough
Project manager: Mike Harman (Life Lines Editorial Services)
Development editor: Joan Miller
Copyeditor: Tanya Solomons
Proofreader: Catherine Dakin
Cover designer: Gordon MacGilp
Cover illustrator: Ann Paganuzzi
Typesetter: Ken Vail Graphic Design
Illustrators: Ann Paganuzzi and QBS Learning
Production controller: Lyndsey Rogers
Printed in India by Multivista Global Pvt. Ltd.

With thanks to the following teachers and schools for reviewing materials in development: Calcutta International School; Hawar International School; Melissa Brobst, International School of Budapest; Rafaella Alexandrou, Pascal Primary Lefkosia; Maria Biglikoudi, Georgia Keravnou, Sotiria Leonidou and Niki Tzorzis, Pascal Primary School Lemessos; Taman Rama Intercultural School, Bali.

MIX
Paper | Supporting
responsible forestry
FSC
www.fsc.org
FSC™ C007454

This book is produced from independently certified FSC™ paper to ensure responsible forest management.

For more information visit:
www.harpercollins.co.uk/green

Contents

Number

Geometry and Measure

Statistics and Probability

How to use this book

This book is used during the part of a lesson when you practise the mathematical ideas you have just been taught.

- An **objective** explains what you should know, or be able to do, by the end of the lesson.

You will need
- Shows the things you need to use to answer some of the questions.

There is a page of practice questions for each lesson, with three different types of questions:

1 Some question numbers are written on a **circle**. These questions may be **easier**. They may also practise ideas you have learned before. These questions will help you answer the rest of the questions on the page.

2 Some question numbers are written on a **triangle**. These questions help you better understand the ideas you have just been taught.

3 Some question numbers are written on a **square**. These are a little **harder** and make you think.

You won't always have to answer all the questions on the page. Your teacher will tell you which questions to answer.

HINT

Circle the question numbers your teacher tells you to answer.

Date: _____

At the bottom of the page there is room to write the date you completed the work. If it took you longer than 1 day, write all of the dates you worked on the page.

Self-assessment

Once you've answered the questions, think about how easy or hard you find the ideas. Circle the face that describes you best.

☺ I can do this.

😐 I'm getting there.

☹ I need some help.

Lesson 1: **Counting objects**

• Count up to 10 objects

Count the ice-cream cones.

1

2

Count the sweets.

3

4

5

6

7 One group of cakes has fewer in it than the others.
Circle the group with fewer cakes.

Date: _____

Number

Lesson 2: **Counting on and back in ones**

* Count forwards and backwards in ones to 10

1 Count on 1 from the star number. Colour the number.

1	2	☆3	4	5	6	7	8	9	10

2 Count back 1 from the star number. Colour the number.

1	2	3	4	5	☆6	7	8	9	10

3 Count forwards in ones. Which number goes in the star?

1		3	4		☆	7		9	

4 Count forwards in ones. Which number goes in the star?

1	2	3	4		6		8	☆	10

5 Count backwards in ones. Which number goes in the star?

10		8	☆	6		4		2	1

6 Count backwards in ones. Which number goes in the star?

	9		7	6		☆	3	2	

7 Pick a number from 1 to 9. Write it here. ☐

Count on one. What number are you on now? ☐

8 Pick a number from 1 to 9. Write it here. ☐

Count back one. What number are you on now? ☐

Date: _____

Number

Lesson 3: **Sequences of objects**

• Recognise numbers as patterns

1 Match each number to the correct pattern.

5   3   1   4

Tick the number that matches.

2 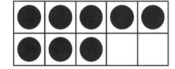   8 ☐   5 ☐   10 ☐

3 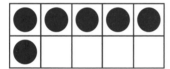   7 ☐   6 ☐   5 ☐

4    4 ☐   5 ☐   6 ☐

5 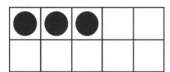   3 ☐   2 ☐   4 ☐

6 Show two different ways to show 5 in a pattern.

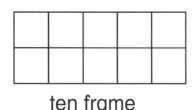

ten frame   dice

Date: _____

Lesson 4: **Estimating to 10**

- Estimate how many objects there are and check this by counting

1 Estimate, then count.

Estimate ☐   Count ☐

Estimate, then count. Use the stars to help you estimate.

2

Estimate ☐   Count ☐

3

Estimate ☐   Count ☐

4

Estimate ☐   Count ☐

5 Estimate, then count.

Estimate ☐   Count ☐

Date: _____

9

Lesson 1: **Estimating to 20**

- Estimate amounts of objects to 20

1 Estimate: More or less than 10? Draw a ring around your answer.

more	less

2 Estimate: More or less than 10? Draw a ring around your answer.

more	less

3 Estimate, then count.

a
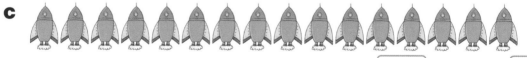

Estimate [ ]   Amount [ ]

b

Estimate [ ]   Amount [ ]

c

Estimate [ ]   Amount [ ]

d

Estimate [ ]   Amount [ ]

4 Tick the group that you estimate to contain 17 stars.

a    **b**    **c**

Date: _____

Number

Lesson 2: **Counting in twos**

- Count on in twos from any number to 20
- Notice and describe number patterns

1 Count on in twos from the star number to fill in the missing numbers.

a

| 0 | 1 | ☆2 | 3 | | 5 | | 7 | | 9 | |

b

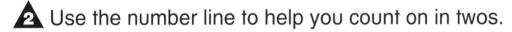

| 0 | 1 | 2 | ☆3 | 4 | | 6 | | 8 | | 10 |

2 Use the number line to help you count on in twos.

0 1 2 3 4 5 6 7 8 9 10 11 12 13 14 15 16 17 18 19 20

a

6 ☐ ☐ ☐ ☐ 16

b

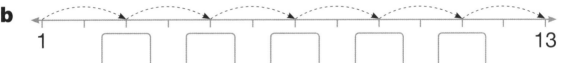

1 ☐ ☐ ☐ ☐ ☐ 13

c

12 ☐ ☐ ☐ 20

d

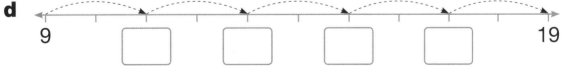

9 ☐ ☐ ☐ ☐ 19

3 Count on in twos from the circled number. What number goes in the box on the number line?

④ ☐

Date: _____ ☺ ☺ ☹

Lesson 3: **Odd and even numbers**

Number

• Recognise odd and even numbers to 20

1 How many flowers? ☐ Circle pairs to help you.

Is the number odd or even? ☐

2 How many leaves? ☐ Circle pairs to help you.

Is the number odd or even? ☐

Use the number line to help you.

0 1 2 3 4 5 6 7 8 9 10

3 Write the odd numbers. ☐☐☐☐☐

4 Write the even numbers. ☐☐☐☐☐

5 Circle the odd numbers. Use the number line to help you.

13   7   12   14   17   3   15   18

20   16   19   1   6   10   5   11

Date: _____

Lesson 4: **Counting in tens**

> • Count on and back in tens

1 Count on from 4 to find 10 **more**. ☐

0 1 2 3 4 5 6 7 8 9 10 11 12 13 14 15 16 17 18 19 20

2 Count back from 12 to find 10 **less**. ☐

0 1 2 3 4 5 6 7 8 9 10 11 12 13 14 15 16 17 18 19 20

3 Count on or back in tens to find the answers.

   a 10 **more** than 7 is: ☐   **b** 10 **more** than 9 is: ☐

   c 10 **more** than 0 is: ☐   **d** 10 **less** than 15 is: ☐

   e 10 **less** than 20 is: ☐   **f** 10 **less** than 11 is: ☐

4 Count on or back in tens to find the answers.

   a 6 monkeys are swinging from the trees.
     10 more monkeys join in.
     How many monkeys are there now? ☐

   b 19 lions are hunting in the jungle.
     10 of them hide in a cave.
     How many lions are left? ☐

Date: _____

Number

Lesson 1: **Counting to 10**

• Count to 10

1 Trace the line and count on.

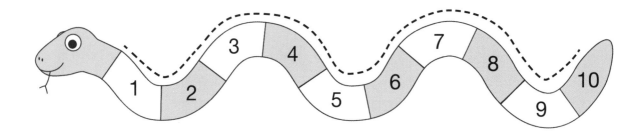

2 Trace the line and count on.

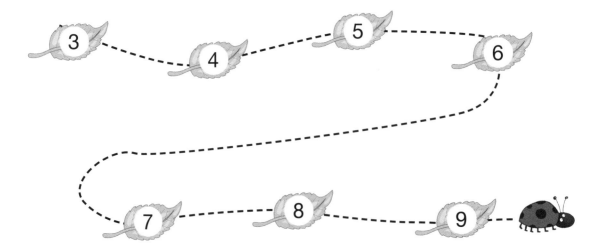

3 Trace the line and count on.

Date: _____

Lesson 2: **Reading numbers to 10**

- Read the numbers 1 to 10

1	2	3	4	5	6	7	8	9	10

a Circle how many eyes you have.

b Shade how many noses you have.

c Draw a triangle around how many fingers you have on one hand.

2 Count and then draw a line from each group to the matching number.

1	2	3	4	5	6	7	8	9	10

3 Draw flowers to match the numbers.

10	7

Date: _____

Number

Lesson 3: **Writing numbers to 10**

- Write the numbers 1 to 10

1 Trace over the numbers.

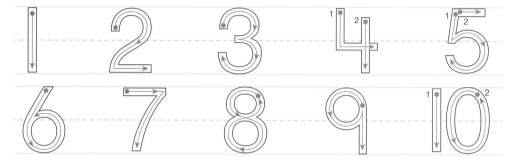

2 Copy the numbers.

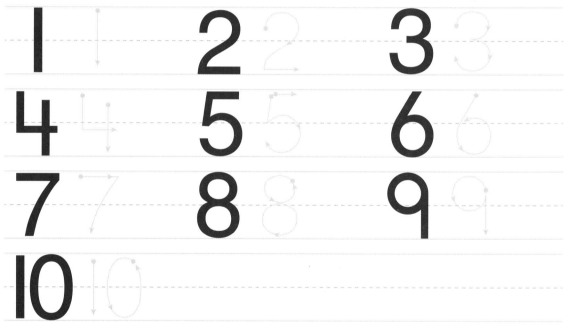

3 Write your age. ☐

4 Fill in the missing numbers.

Date: _____

☺ 😐 ☹

Lesson 4: **How many?**

- Count up to 10 objects

1 How many leaves? ☐ Circle each one as you count it.

 1 2 3 4 5 6 7 8 9 10

2 How many stars? ☐ Circle each one as you count it.

3 How many planets? ☐ Circle each one as you count it.

4 How many rockets? ☐ Circle each one as you count it.

5 How many flowers? ☐ Circle each one as you count it.

Date: _____

Number

Lesson 1: **Counting to 20**

- Count forwards and backwards from any given number to 20

1 Trace the line and count on.

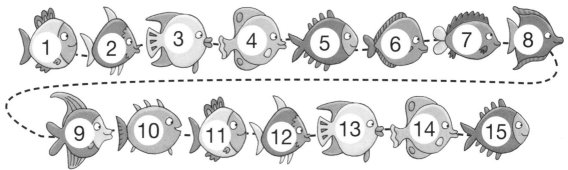

2 Trace the line and count on.

3 Trace the line and count back.

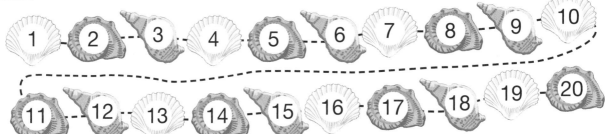

4 Trace the line and count on. Write in the starting number.

Date: _____

Number

Lesson 2: **Reading numbers to 20**

• Read numbers to 20

1 Draw lines to match the numerals to the words.

11	fifteen
20	twenty
15	eleven

2 Count how many, then draw lines to match the numeral and word to the matching amount.

12   fourteen

19   twelve

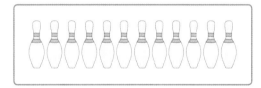

14   nineteen

3 Circle the words and numerals that are matched to the **wrong** amount.

a    11   twelve

b    13   fifteen

Date: _____

Lesson 3: **Writing numbers to 20**

• Write numerals and number words to 20

1 Write the matching numerals and words.

a twenty ☐   **b** 16   s _ xt _ _ n

c eleven ☐   **d** 19   ni _ _ t _ _ n

2 Write the matching numeral or word.

a twelve ☐   **b** 9   _____

c fourteen ☐   **d** 15   _____

e ten ☐   **f** 13   _____

g eight ☐   **h** 7   _____

3 Fill in the missing numbers then write each number as a word.

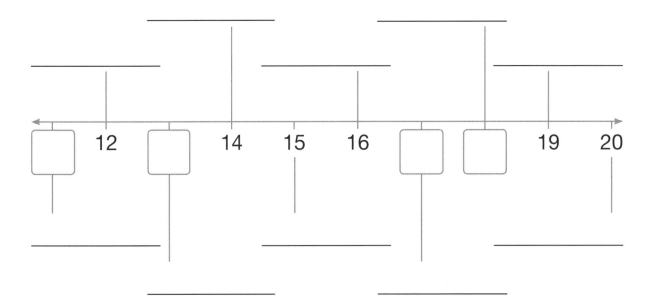

Number

Date: _____

Number

Lesson 4: **Counting and labelling objects to 20**

- Count up to 20 objects accurately
- Label amounts to 20

1 Count the bees. ☐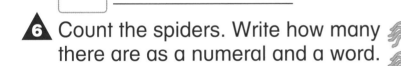

2 Count the caterpillars. ☐

3 Count the butterflies. Write how many there are as a numeral and a word.

☐ _____

4 Count the beetles. Write how many there are as a numeral and a word.

☐ _____

5 Count the snails. Write how many there are as a numeral and a word.

☐ _____

6 Count the spiders. Write how many there are as a numeral and a word.

☐ _____

7 Draw 19 flowers. Write the word for 19.

_____

8 Draw 14 leaves. Write the word for 14.

_____

Date: _____

Number

Lesson 1: **Combining sets**

• Combine two sets to find how many altogether

1   □

2   □

3 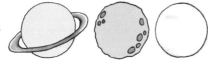 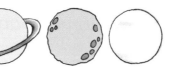 □

4   □

5   □

6 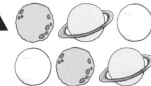 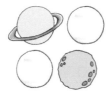 □

7 Draw a set of 5 rockets. Draw a set of 4 rockets. Add them together.

5	4

□

Date: _____

22

Lesson 2: **Part–whole diagrams**

• Use part–whole diagrams to combine sets of objects

How many altogether?

1

2

3

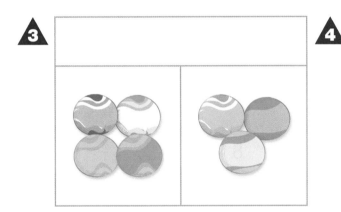

4

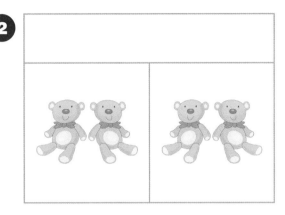

5

6

Date: _____

Number

Lesson 3: **Writing addition number sentences**

- Use + and = in an addition number sentence
- Use groups of objects to make an addition number sentence

Number

1 Count how many altogether to find the total.

a ⬤⬤⬤⬤ + ⬤⬤ = ☐

b ▢▢▢▢▢ + ▢▢▢ = ☐

2 Complete the number sentences to match the groups of objects.

a △△△△△ △△   5 + ☐ = ☐

b ⬭⬭⬭⬭ ⬭⬭⬭   ☐ + 3 = ☐

c ▭▭▭▭ ▭   8 + ☐ = ☐

d ★★★★ ★★★★   ☐ + ☐ = 8

3 Write number sentences to match the groups of objects.

a ▢▢▢▢▢ ▢▢▢

_____

b ⬤⬤⬤⬤ ⬤⬤⬤

_____

Date: _____

Number

Lesson 4: Using sets of objects to solve additions

• Solve additions by combining groups of objects

Use the part–whole diagram to find the answer to each addition.

1 2 + 1 = ☐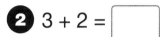

2 3 + 2 = ☐

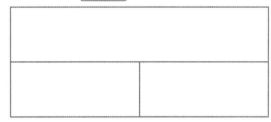

3 6 + 1 = ☐

4 3 + 3 = ☐

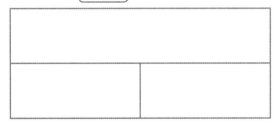

5 4 + 3 = ☐

6 6 + 2 = ☐

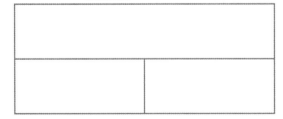
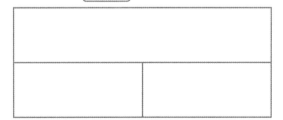

7 7 + 2 = ☐

8 9 + 0 = ☐

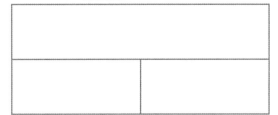

Date: _____ ☺ 😐 ☹

Number

Lesson 1: **Adding more**

• Add two amounts by counting on with objects

Count on from the starting amount.

1 Add 1 more.

2 Add 2 more.

3 Add 2 more.

4 Add 3 more.

5 Add 1 more.

6 Add 4 more.

7 Leo has 3 cats and Susie has 4 cats.

How many cats is that altogether?

8 There are 6 mice. 2 more mice join them.

How many mice are there now?

Date: _____

Number

Lesson 2: **Adding more to solve additions**

• Solve additions by counting on more objects

You will need

• 10 interlocking cubes (Challenge 1 only)

1 Use cubes to help you to count on.

a 4 + 2 = ☐

b 5 + 1 = ☐

c 5 + 3 = ☐

d 2 + 3 = ☐

2 Draw the extra shapes to count on.

a 7 + 1 = ☐

| 1 | 2 | 3 | 4 | 5 | 6 | 7 |

b 4 + 3 = ☐   ① ② ③ ④

c 6 + 2 = ☐

| 1 | 2 | 3 | 4 | 5 | 6 |

d 8 + 2 = ☐

△1 △2 △3 △4 △5 △6 △7 △8

3 Draw dots to help you count on to find each answer. Remember to put the larger number first.

a 2 + 5 = ☐   ☐

b 3 + 6 = ☐   ☐

Date: _____

Number

Lesson 3: **Adding more with a number track**

• Use a number track to count on to solve additions

1 Put your finger on the starting number then count on.
Circle the number you land on.

a 7 + 2 =

b 5 + 3 =

2 Count on along the number track from the larger number to find the totals.

1 2 3 4 5 6 7 8 9 10

a 7 + 3 = ☐   **b** 2 + 5 = ☐

c 3 + 1 = ☐   **d** 4 + 5 = ☐

3 Use the number track in **2** to count on from the larger number to find the totals. Write a number sentence to match.

a Jamaal has 5 marbles in a bag. He buys 3 more.
How many marbles does he have now?

☐ + ☐ = ☐

b Shaia is playing with 2 friends. 4 more friends join them.
How many friends is Shaia playing with now?

☐ + ☐ = ☐

Date: _____

Number

Lesson 4: **Adding more with a number line**

• Use a number line to count on to solve additions

1 Draw the jumps on the number line and circle the number you land on.

a $3 + 2 =$

0   1   2   3   4   5

b $2 + 2 =$

0   1   2   3   4   5

2 Draw jumps on the number line to count on and draw a ring around the total. Remember to start on the larger number.

a $3 + 5 = \boxed{}$

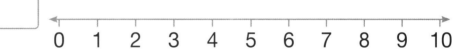

0   1   2   3   4   5   6   7   8   9   10

b $1 + 8 = \boxed{}$

0   1   2   3   4   5   6   7   8   9   10

c $4 + 2 = \boxed{}$

0   1   2   3   4   5   6   7   8   9   10

d $7 + 3 = \boxed{}$

0   1   2   3   4   5   6   7   8   9   10

3 Write a number sentence to match each number line.

a

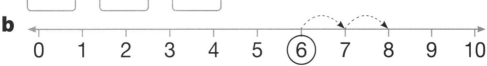

0   1   2   3   ④   5   6   7   8   9   10

$\boxed{} + \boxed{} = \boxed{}$

b

0   1   2   3   4   5   ⑥   7   8   9   10

$\boxed{} + \boxed{} = \boxed{}$

Date: _____

Number

Lesson 1: **Taking away objects**

• Take objects from a set and re-count to find how many are left

You will need
• 10 counters

1 Take away 1. How many are left? ☐

① ② ③ ④

2 Take away 2. How many are left? ☐

① ② ③ ④ ⑤

3 Count out 6 counters. Now take away 3.

How many are left? ☐

4 Count out 7 counters. Now take away 3.

How many are left? ☐

5 Count out 5 counters. Now take away 4.

How many are left? ☐

6 Count out 10 counters. Now take away 6.

How many are left? ☐

Count out counters to match the starting amount, then take some away to find the answers.

7 There are 9 ants on the floor. 4 run away.

How many are left? ☐

8 There are 8 bananas on the tree. 5 fall off.

How many bananas are left on the tree? ☐

Date: _____

Lesson 2: **Taking away to solve subtractions**

• Solve subtractions by taking away objects from a group

1 Take away. How many are left?

a 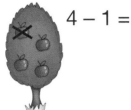 $4 - 1 =$ ☐

b 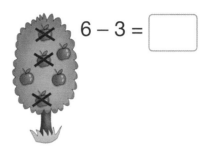 $6 - 3 =$ ☐

2 Take away the number shown.

a 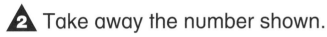 **3** How many are left? ☐

b  **5** How many are left? ☐

3 Cross out the buttons. Then write the answer.

a

$6 - 2 =$ ☐

b

$10 - 4 =$ ☐

4 Write the subtraction and complete the answer.

a
☐ – ☐ = ☐

b
☐ – ☐ = ☐

Date: _____

Number

Lesson 3: **Taking away with part–whole diagrams**

• Use a part–whole diagram to take away

1 Draw how many counters are left in each part–whole diagram.

a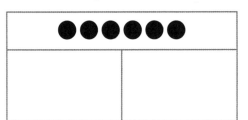

b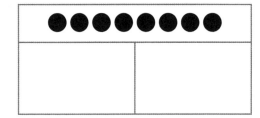

2 Draw the counters to complete each part–whole diagram.

a Take away 3.

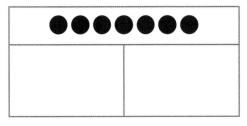

b Take away 2.

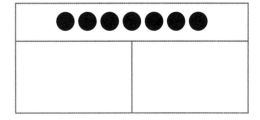

c Take away 1.

d Take away 4.

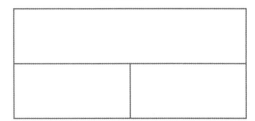

3 Fill in the part–whole diagrams.

a $9 - 3 = \boxed{}$

b $10 - 5 = \boxed{}$

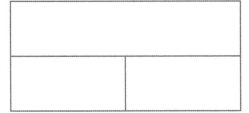

Date: _____

Number

Lesson 4: **Solving subtractions with part–whole diagrams**

• Use a part–whole diagram to solve subtractions

Use the part–whole diagram to find the answer to each subtraction.

1 $2 - 1 =$ ☐

2 $5 - 2 =$ ☐

3 $7 - 1 =$ ☐

4 $6 - 2 =$ ☐

5 $8 - 3 =$ ☐

6 $9 - 4 =$ ☐

7 $9 - 8 =$ ☐

8 $10 - 7 =$ ☐

Date: _____

☺ ☹ ☺

33

Number

Lesson 1: **Counting back in ones to subtract**

• Count back in ones to subtract an amount

Colour in the number of shapes that you are subtracting.
Then count back along the coloured shapes to find the answer.

1 a Subtract 1. ① ② ③ ④ ⑤ ☐

b Subtract 2. | 1 | 2 | 3 | 4 | 5 | ☐

2 a Subtract 2.

b Subtract 3.

| 1 | 2 | 3 | 4 | 5 | 6 | 7 | ☐

c Subtract 3. ① ② ③ ④ ⑤ ⑥ ☐

d Subtract 4.

| 1 | 2 | 3 | 4 | 5 | 6 | 7 | 8 | 9 | 10 | ☐

3 Draw the starting number of circles. Colour in the number of circles that you are subtracting. Then count back along the coloured circles to find the answer.

a Starting number: 9   Subtract 5

☐   ☐

b Starting number: 10   Subtract 7

☐   ☐

Date: _____

Lesson 2: **Subtracting on a number track**

> • Solve subtractions by counting back on a number track

1 Solve these subtractions by counting back from 5.

| 0 | 1 | 2 | 3 | 4 | 5 |

a 5 – 3 = ☐   **b** 5 – 4 = ☐

2 Solve these subtractions by counting back.

0 1 2 3 4 5 6 7 8 9 10

a 10 – 4 = ☐   **b** 6 – 3 = ☐

c 9 – 6 = ☐   **d** 4 – 4 = ☐

e 8 – 6 = ☐   **f** 7 – 3 = ☐

3 Draw a number track then count back to solve
the subtraction.

8 – 6 = ☐

Date: _____

Number

Number

Lesson 3: **Subtracting on a number line**

• Solve subtractions by counting back on a number line

1 Use the number lines to solve these subtractions.

a 7 – 2 = ☐

0 1 2 3 4 5 6 7

b 9 – 3 = ☐

0 1 2 3 4 5 6 7 8 9

2 Use the number lines to solve these subtractions.

a 10 – 5 = ☐

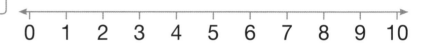
0 1 2 3 4 5 6 7 8 9 10

b 8 – 4 = ☐

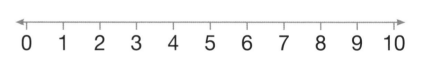
0 1 2 3 4 5 6 7 8 9 10

c 4 – 4 = ☐

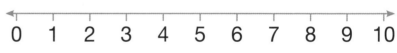
0 1 2 3 4 5 6 7 8 9 10

d 7 – 6 = ☐

0 1 2 3 4 5 6 7 8 9 10

3 Draw number lines to solve these subtractions.

a

9 – 7 = ☐

b

6 – 5 = ☐

Date: _____

Lesson 4: **Counting back to solve subtractions**

- Use mental strategies to count back to solve subtractions

Count back in your head to solve the subtractions.

Then use the number line to check your answers.

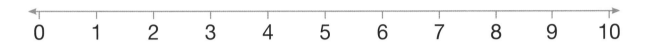

0   1   2   3   4   5   6   7   8   9   10

1 10 – 2 = ☐

2 10 – 3 = ☐

3 10 – 5 = ☐

4 10 – 1 = ☐

5 5 – 2 = ☐

6 3 – 1 = ☐

7 8 – 6 = ☐

8 9 – 9 = ☐

9 10 – 7 = ☐

10 4 – 3 = ☐

11 Write two different subtractions that have 3 as the answer.

3 ☐ ☐

12 Write two different subtractions that have 5 as the answer.

5 ☐ ☐

Date: _____

Number

37

Number

Lesson 1: **Finding the difference**

> • Find the difference between two amounts

1 Count how many extra cubes there are in the longer line to find the difference.

a  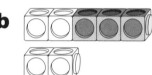 **b**

2 Circle the extra animals in each line. Count them to find the difference.

a

b

c

d

3 Draw two lines of circles to find the difference between these two amounts.

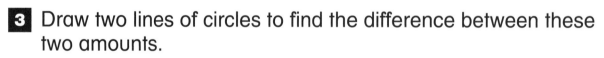

8 and 3

Date: _____

Lesson 2: **Subtracting by finding the difference**

• Solve subtractions by finding the difference

1 Cross out the extra sweets to find the difference.

a 6 – 4 = ☐

b 8 – 7 = ☐

2 Draw counters to find the difference.

a 8 – 5 = ☐

b 4 – 2 = ☐

c 9 – 6 = ☐

d 10 – 2 = ☐

3 Write subtraction number sentences to match these cakes.

a
   ☐ – ☐ = ☐

b
   ☐ – ☐ = ☐

Date: _____   ☺ ☺ ☹

Number

Lesson 3: **Finding the difference on a number line**

• Find the difference between two numbers on a number line

Use the number lines to find the difference between each pair of numbers.

1 **a** 3 and 9 [ ]

1 2 ③ 4 5 6 7 8 ⑨ 10

b 4 and 6 [ ]

1 2 3 ④ 5 ⑥ 7 8 9 10

2 **a** 3 and 5 [ ]

0 1 2 3 4 5 6 7 8 9 10

b 1 and 6 [ ]

0 1 2 3 4 5 6 7 8 9 10

c 7 and 9 [ ]

0 1 2 3 4 5 6 7 8 9 10

d 4 and 10 [ ]

0 1 2 3 4 5 6 7 8 9 10

3 Write a pair of numbers that have a difference of 3. Use the number line to help you.

[ ] and [ ]

0 1 2 3 4 5 6 7 8 9 10

4 Write a pair of numbers that have a difference of 5. Use the number line to help you.

[ ] and [ ]

0 1 2 3 4 5 6 7 8 9 10

Date: _____

Number

Lesson 4: **Subtraction as difference on a number line**

• Solve subtractions by finding the difference on a number line

1 Use the number line to find difference.

a $5 - 2 = \boxed{}$

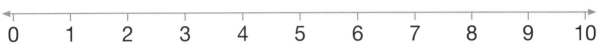

b $6 - 4 = \boxed{}$

2 Use the number line to find difference.

a $9 - 8 = \boxed{}$ **b** $7 - 0 = \boxed{}$

c $7 - 4 = \boxed{}$ **d** $10 - 1 = \boxed{}$

3 Write a subtraction number sentence to match this problem, then find the difference on the number line to solve it.

Bella picked 4 flowers. Rania picked 8 flowers. What is the difference between the amounts of flowers they picked?

$\boxed{} - \boxed{} = \boxed{}$

Date: _____

41

Lesson 1: **Making 10**

• Find pairs of numbers that total 10

1 How many blocks do you need to colour in to make 10 coloured blocks?

a  9 and [ ]

b  7 and [ ]

2 Draw sweets in the empty bag to make 10 sweets altogether. Write how many sweets are in each bag.

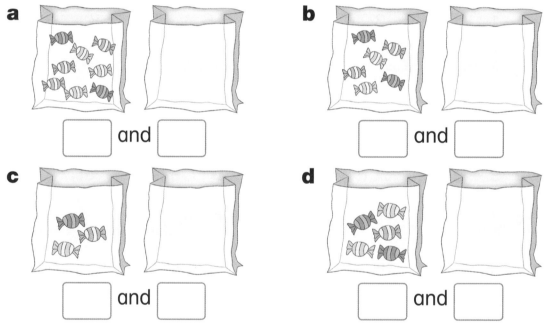

a [ ] and [ ]

b [ ] and [ ]

c [ ] and [ ]

d [ ] and [ ]

3 Draw dots in the pairs of boxes to show three different ways to make 10.

a [ ]
[ ]

b [ ]
[ ]

c [ ]
[ ]

Date: _____

Lesson 2: **Addition and subtraction facts for 10**

> • Find addition and subtraction facts for 10

1 Write an addition number sentence.

10	
9	1

☐ + ☐ = 10

2 Write a subtraction number sentence.

10	
8	2

10 − ☐ = ☐

3 Write one addition and one subtraction number sentence.

a

10	
6	4

☐ + ☐ = ☐

☐ − ☐ = ☐

b

10	
5	5

☐ + ☐ = ☐

☐ − ☐ = ☐

c

10	
7	3

☐ + ☐ = ☐

☐ − ☐ = ☐

d

10	
10	0

☐ + ☐ = ☐

☐ − ☐ = ☐

4 Fill in the part–whole diagram with two numbers that total 10. Write two additions and two subtractions to match.

10	

☐ + ☐ = ☐   ☐ − ☐ = ☐

☐ + ☐ = ☐   ☐ − ☐ = ☐

Date: _____

Number

Lesson 3: **Making numbers to 10**

• Find pairs of numbers that make totals to 10

Use cubes to help you find pairs of numbers that total 5, 7 and 9.

You will need
• 10 interlocking cubes

1

5

5 and 0　　　0 and ▢

4 and ▢　　　1 and ▢

▢ and 2　　　3 and ▢

2

7

7 and ▢　　　0 and ▢

6 and ▢　　　1 and ▢

▢ and ▢　　　2 and ▢

4 and ▢　　　▢ and ▢

3

9

▢ and ▢　　　0 and 9

8 and ▢　　　1 and ▢

▢ and 2　　　▢ and 7

6 and 3　　　▢ and 6

▢ and 4　　　4 and 5

Date: _____

Lesson 4: **Addition and subtraction facts to 10**

• Find addition and subtraction facts to 10

1 Write one addition and one subtraction for the complement of 4.

4	
3	1

$\boxed{} + \boxed{} = 4$

$4 - \boxed{} = \boxed{}$

2 Write one addition and one subtraction for the complement of 5.

5	
2	3

$\boxed{} + \boxed{} = 5$

$5 - \boxed{} = \boxed{}$

3 Write two additions and two subtractions for the complement of 7.

7	
5	2

$\boxed{} + \boxed{} = 7$

$\boxed{} + \boxed{} = 7$

$7 - \boxed{} = \boxed{}$

$7 - \boxed{} = \boxed{}$

4 Write two additions and two subtractions for the complement of 9.

9	
6	3

$\boxed{} + \boxed{} = 9$

$\boxed{} + \boxed{} = 9$

$9 - \boxed{} = \boxed{}$

$9 - \boxed{} = \boxed{}$

5 Circle the number sentences that do **not** match the complement of 8.

8	
5	3

$3 + 5 = 8$ $8 - 5 = 3$ $3 - 8 = 5$

$5 + 3 = 8$ $8 + 3 = 5$ $8 - 3 = 5$

Date: _____

Lesson 1: **Estimating an answer**

• Estimate the answer to an addition or subtraction to 10

1 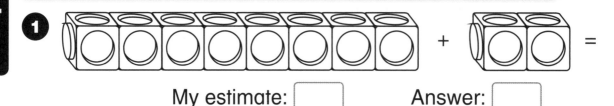   + =

My estimate: ☐   Answer: ☐

2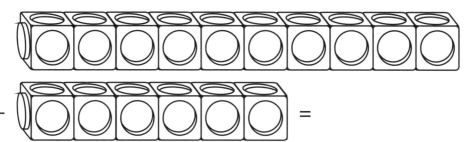

− =

My estimate: ☐   Answer: ☐

3 3 + 2 =   My estimate: ☐   Answer: ☐

4 7 + 3 =   My estimate: ☐   Answer: ☐

5 8 − 6 =   My estimate: ☐   Answer: ☐

6 5 − 4 =   My estimate: ☐   Answer: ☐

7 5 + 4 =   My estimate: ☐   Answer: ☐

Tick the best estimates for each answer. You may choose more than one.

8 6 + 3 = | 8 | | 2 | | nearly 10 | | under 5 | | 4 |

9 6 − 3 = | 10 | | less than 5 | | more than 6 | | 2 | | 6 |

Date: _____

Lesson 2: **Choosing how to solve an addition**

- Use different strategies to solve additions

Number

Choose a method or piece of equipment to solve each addition.

You could count on or combine two sets and you could use:
- cubes
- a part–whole diagram
- counters
- a number line.

1 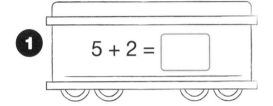 5 + 2 = ☐

2 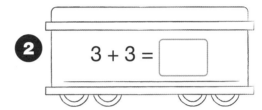 3 + 3 = ☐

3 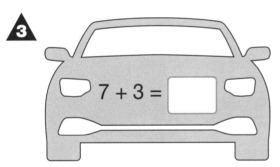 7 + 3 = ☐

4 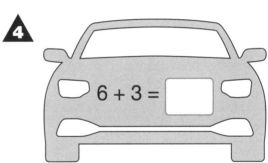 6 + 3 = ☐

5 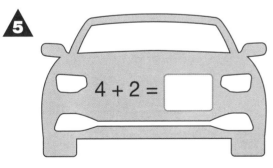 4 + 2 = ☐

6 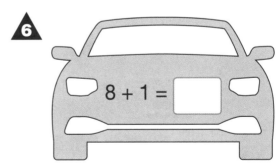 8 + 1 = ☐

7 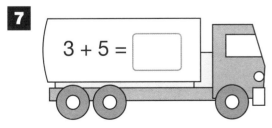 3 + 5 = ☐

8 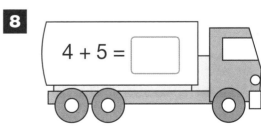 4 + 5 = ☐

Date: _____

Number

Lesson 3: **Choosing how to solve a subtraction**

• Use different strategies to solve subtractions

Choose a method or piece of equipment to solve each subtraction.

You could take away, count back or find the difference and you could use:

- cubes
- a part–whole diagram

- counters
- a number line.

1 $7 - 2 =$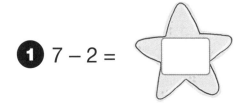

2 $9 - 1 =$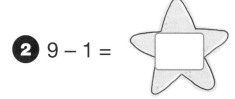

3 $10 - 5 =$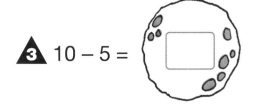

4 $5 - 4 =$

5 $8 - 6 =$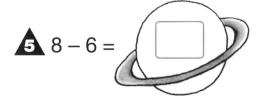

6 $9 - 3 =$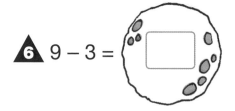

7 Write your own subtraction and solve it using two different methods.

What methods did you use?

_____

Date: _____

Lesson 4: **Addition and subtraction in real life**

- Relate addition and subtraction to real-life situations
- Use a chosen strategy to solve additions and subtractions

Number

1 Do you need to add or subtract to solve each of these problems? Draw a ring around your answer.

a I picked 4 apples. Then I picked 1 more. How many apples do I have altogether?

| + | – |

b Hiro had 8 cherries in his bowl. He ate 3 cherries. How many cherries were left?

| + | – |

2 Write the number sentences for these additions and subtractions.

a Jayesh had 6 toy cars. He was given 2 more. How many does he have now?

☐ ◯ ☐ ◯ ☐

b Olivia picks 7 oranges. Luca picks 3 oranges. How many oranges do they have altogether?

☐ ◯ ☐ ◯ ☐

c There were 8 people on a bus. 4 got off. How many people are on the bus now?

☐ ◯ ☐ ◯ ☐

d Amina had 5 sweets. She dropped 2 on the floor. How many did she have left?

☐ ◯ ☐ ◯ ☐

3 Write a real-life addition and solve it. _____

_____

_____ ☐ ◯ ☐ ◯ ☐

Date: _____

Lesson 1: **Addition facts to 20**

> • Add 1-digit and 2-digit numbers to 20 by counting on

1 Use the stars to count on.

    a 12 + ⭐⭐⭐ = ☐   **b** 11 + ⭐⭐⭐ = ☐

    c 15 + ⭐⭐ = ☐   **d** 17 + ⭐⭐ = ☐

    e 11 + ⭐⭐⭐⭐⭐ = ☐   **f** 13 + ⭐⭐⭐⭐⭐ = ☐

2 Use the number line to count on.

```
0  1  2  3  4  5  6  7  8  9  10 11 12 13 14 15 16 17 18 19 20
```

    a 17 + 3 = ☐   **b** 12 + 6 = ☐

    c 16 + 2 = ☐   **d** 11 + 6 = ☐

    e 14 + 6 = ☐   **f** 12 + 5 = ☐

    g 15 + 4 = ☐   **h** 13 + 7 = ☐

3 Start with the larger number, then count on in your head.

    a 7 + 11 = ☐   **b** 5 + 14 = ☐

    c 6 + 13 = ☐   **d** 4 + 16 = ☐

Date: _____ 😊 😐 ☹

Lesson 2: **Making 10 and adding more**

• Use number pairs that total 10 to add

1 First make 10, then add what's left over.

a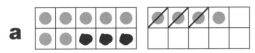

b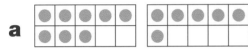

2 First make 10, then add what's left over.

a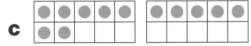

☐ + ☐ = ☐

b

☐ + ☐ = ☐

c

☐ + ☐ = ☐

d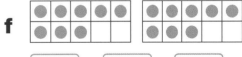

☐ + ☐ = ☐

e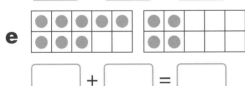

☐ + ☐ = ☐

f

☐ + ☐ = ☐

3 Complete the additions, making 10 first.

a 5 + 7 | 5 | + | 5 | + | 2 | = | 12 |

b 7 + 7   ☐ + ☐ + ☐ = ☐

c 6 + 5   ☐ + ☐ + ☐ = ☐

d 8 + 7   ☐ + ☐ + ☐ = ☐

Date: _____

Lesson 3: **Near doubles**

• Solve additions by finding near doubles

1 Complete the double, then the 'double + 1' calculation.

a 2 + 2 = ☐   so 2 + 3 = ☐

b 4 + 4 = ☐   so 4 + 5 = ☐

2 Use doubling to work out the answers.

a 3 + 4

Double ☐3☐ = ☐6☐

☐6☐ ⊕ ☐1☐ = ☐7☐

b 6 + 7

Double ☐ = ☐

☐ ◯ ☐ = ☐

c 8 + 9

Double ☐ = ☐

☐ ◯ ☐ = ☐

d 9 + 10

Double ☐ = ☐

☐ ◯ ☐ = ☐

3 Use doubling to work out the answers.

a 7 + 9

Double ☐ = ☐

☐ ◯ ☐ = ☐

b 6 + 8

Double ☐ = ☐

☐ ◯ ☐ = ☐

Date: _____

Number

Lesson 4: **Addition and estimation to 20**

- Use different strategies to solve additions
- Estimate an answer for an addition

1 Estimate, then put the 2-digit number first and count on. There were 12 cars in the car park. 4 more drove in. How many cars are there now?

Estimate = ☐　　　☐ + ☐ = ☐

2 Estimate, then make a 10 and add more. Zoe fed 8 parakeets and 3 parrots. How many birds did she feed?

Estimate = ☐　　　☐ + ☐ = ☐

Estimate, then choose a strategy to solve the additions.

3 9 children line up for a snack. 4 more children join them. How many children are there altogether?

Estimate = ☐　　　☐ + ☐ = ☐

4 There were 14 people on a bus. At the next stop 5 more people got on the bus. How many people are now on the bus?

Estimate = ☐　　　☐ + ☐ = ☐

5 Ben grew 9 sunflowers. Jack grew 10 sunflowers. How many sunflowers did they grow altogether?

Estimate = ☐　　　☐ + ☐ = ☐

6 **a** I had 9 marbles in my pocket. My friend gave me 5 more marbles. How many marbles do I have now?

Estimate = ☐　　　☐ + ☐ = ☐

b How did you work out the answer?

_____

Date: _____　　☺ 😐 ☹

Number

Lesson 1: **Subtraction facts to 20**

• Subtract a 1-digit number from a 2-digit number to 20

1 Cross out the ones to help you solve the subtraction.

a 17 – 2 = ⬚   **b** 15 – 4 = ⬚

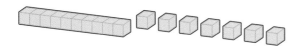 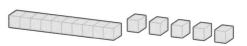

2 Use what you know to solve each subtraction. Then use the number line to check your answer.

0 1 2 3 4 5 6 7 8 9 10 11 12 13 14 15 16 17 18 19 20

a 16 – 3 = ?

6 – 3 = ⬚

So 16 – 3 = ⬚

b 18 – 6 = ?

8 – 6 = ⬚

So 18 – 6 = ⬚

c 13 – 2 = ?

3 – 2 = ⬚

So 13 – 2 = ⬚

d 19 – 5 = ?

9 – 5 = ⬚

So 19 – 5 = ⬚

3 Count back from the ones number in your head to solve the subtractions.

a 18 – 5 = ⬚   **b** 17 – 7 = ⬚

Date: _____

54

Lesson 2: **Number facts to 20 – part–whole diagrams**

- Find number facts to 20 by using a part–whole diagram

1 Draw dots to make 15.

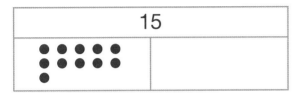

2 Draw dots to make 17.

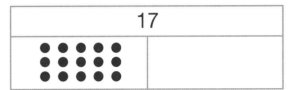

3 Draw dots to find two different ways of making 16.

16	

16	

4 Draw dots to find two different ways of making 20.

20	

20	

5 Draw dots to find two different ways of making 13.

13	

13	

6 Find three different pairs of numbers that total 18.
Fill in the numbers.

18	

18	

18	

Date: _____

Lesson 3: **Number families**

- Find addition and subtraction statements for number families to 20

1 Write additions and subtractions to match this number family.

9	
5	4

☐ + ☐ = 9   ☐ + ☐ = 9

9 – ☐ = ☐   9 – ☐ = ☐

2 Write additions and subtractions to match these number families.

a

8	
6	2

☐ + ☐ = ☐   ☐ + ☐ = ☐

☐ – ☐ = ☐   ☐ – ☐ = ☐

b

7	
4	3

☐ + ☐ = ☐   ☐ + ☐ = ☐

☐ – ☐ = ☐   ☐ – ☐ = ☐

c

12	
7	5

☐ + ☐ = ☐   ☐ + ☐ = ☐

☐ – ☐ = ☐   ☐ – ☐ = ☐

3 Think of a number family and write additions and subtractions to match.

☐ + ☐ = ☐   ☐ + ☐ = ☐

☐ – ☐ = ☐   ☐ – ☐ = ☐

Date: _____

Lesson 4: **Equal statements**

• Find equal statements for numbers to 20

1 Write an equal statement on the other side of the see-saw.

a

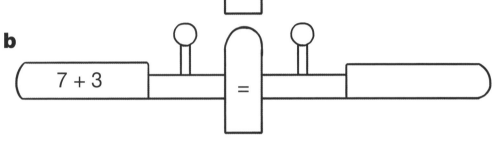

$8 - 2$ $=$

b

$7 + 3$ $=$

2 Write equal statements to match these numbers.

a ⟨14⟩   ☐ + ☐ = ☐ + ☐

b ⟨12⟩   ☐ – ☐ = ☐ – ☐

c ⟨7⟩   ☐ + ☐ = ☐ – ☐

d ⟨13⟩   ☐ + ☐ = ☐ – ☐

3 Pick a number and write equal statements to match.

a ⟨⟩   ☐ ◯ ☐ = ☐ ◯ ☐

b ⟨⟩   ☐ ◯ ☐ = ☐ ◯ ☐

Date: _____

Number

Lesson 1: **Doubling amounts to 5**

• Find and make doubles for amounts to 5

1 Draw the same number of dots again to make double the amount.

a • • •

b •

c • •

d • • • • •

2 Draw the same number of spots on the ladybirds to make double the amount.

a

Double 4 =

b

Double 5 =

c

Double 1 =

d

Double 2 =

e

Double 3 =

f

Double 0 =

3 Draw lines to match each number to its double.
Draw dots to help you if you need to.

 2

 5

 3

 6

 4

 10

Date: _____

Lesson 2: **Doubling amounts to 10**

• Find and make doubles for amounts to 10

1 Count all the cubes to find double the starting amount.

a Double 6 = ☐

b Double 9 = ☐

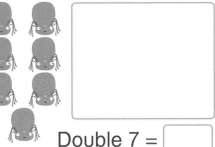

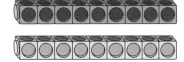
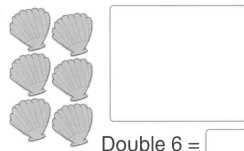

2 Draw the same amount to make double.

a

Double 7 = ☐

b

Double 6 = ☐

c

Double 10 = ☐

d

Double 8 = ☐

3 Circle the number that is double the star number.
Draw dots to help you if you need to.

a **7**   7   10   14   12   11   ☐

b **9**   16   12   14   20   18   ☐

c **10**   18   25   15   20   10   ☐

Date: _____

Number

59

Number

Lesson 3: **Doubling on a number line**

- Find doubles to 10 on a number line

0	1	2	3	4	5	6	7	8	9	10

1 Count along the number track to double the numbers.

a Double 5 = ☐   **b** Double 1 = ☐

c Double 3 = ☐   **d** Double 4 = ☐

2 Count along the number line to double the numbers.

0 1 2 3 4 5 6 7 8 9 10 11 12 13 14 15 16 17 18 19 20

a Double 6 = ☐   **b** Double 8 = ☐

c Double 9 = ☐   **d** Double 0 = ☐

e Double 7 = ☐   **f** Double 10 = ☐

3 Use the number line to check the doubles. Circle the **correct** answers.

0 1 2 3 4 5 6 7 8 9 10 11 12 13 14 15 16 17 18 19 20

Double 2 = 4   Double 7 = 14   Double 9 = 17

Double 5 = 6   Double 8 = 16

Date: _____

☺ 😐 ☹

Lesson 4: **Doubling facts to 10**

- Recall and use doubling facts to 10

1 Double the number and fill in the speech bubble.

a

b

2 Double the number and write it in the box.

a Double 6 = ☐

b Double 9 = ☐

c Double 3 = ☐

d Double 7 = ☐

e Double 4 = ☐

f Double 2 = ☐

g Double 8 = ☐

h Double 0 = ☐

3 Which number has been doubled?

a ☐ → 10

b ☐ → 8

c ☐ → 16

d ☐ → 20

e ☐ → 14

f ☐ → 6

g ☐ → 18

Date: ____________________

☺ 😐 ☹

Number

Lesson 1: **What is money?**

- Know what money is used for
- Recognise features of the money we use

1 Circle the money.

2 Circle any coins that you recognise. If you don't recognise any, draw a coin that you know.

3 Write something that is always on a coin or note.

_____

4 What do we use money for?

_____

_____

Date: _____

Lesson 2: **Recognising coins**

• Recognise the coins we use

You will need
• coloured pencils

Number

1 What is the money that you use called?

_____

2 What colours are the coins you use?

_____

3 Draw three different coins that you use. Think about their colour, shape and any markings that they have on them.

```



```

4 Write some features of the coins you use.

_____

_____

Date: _____

Lesson 3: **Recognising notes**

Number

• Recognise the notes we use

1 What are banknotes made of?

_____

2 What colours are the banknotes you use?

_____

3 Draw a banknote that you use. Think about the colour and what's shown on the note.

4 What are some differences between a note that you use and a note that people in a different country use?

_____

_____

Date: _____

Number

Lesson 4: **Sorting coins and notes**

- Sort the money we use in different ways

You will need
- selection of coins and notes you use
- coloured pencils

1 Put the coins in one group and the notes in another group. Then draw the coins and notes.

Coins	Notes

2 Choose how you would like to sort the coins. Write a heading in each box to show your sorting rule then draw the coins.

_____	_____

3 Choose how you would like to sort the notes. Write a heading in each box to show your sorting rule then draw the notes.

_____	_____

Date: _____ ☺ 😐 ☹

Number

Lesson 1: **Zero**

• Understand that zero means nothing

1 Circle the field that has 0 horses in it.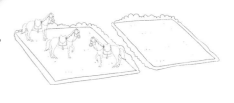

2 Circle the bowl that has 0 fish in it.

3 Draw flowers to match each number.

3		6	

0		8	

4 Draw 1 more .

Draw 0 more 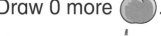.

How many  now? ☐

How many  now? ☐

5 There are 5 frogs in the pond. You add 0 more. How many are there altogether?

☐

6 There are 9 birds on the fence. 0 birds fly away. How many birds are left?

☐

Date: _____

Lesson 2: **Comparing numbers to 10**

- Compare sets of objects and say which has more or less
- Compare numbers to 10

Number

1 Circle the pot with **more** pencils.

2 Circle the pot with **less** paintbrushes.

3 Draw dots in the box to make this correct.

      less   **more**

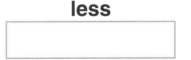

4 Draw dots in the box to make this correct.

      more   **less**

5 Colour the number that is **less**.

6 Colour the number that is **more**.

Circle true or false.

7 8 is more than 5 but less than 10. | true | false |

8 7 is more than 3 but less than 6. | true | false |

Date: _____

Number

Lesson 3: **Ordering numbers to 10**

- Order numbers to 10
- Give a number between two numbers

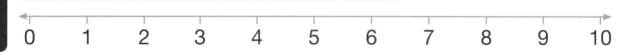

| 0 | 1 | 2 | 3 | 4 | 5 | 6 | 7 | 8 | 9 | 10 |

1 Use the number line to help you fill in the missing numbers.

1   ☐   3   ☐   5   6   ☐   8   9   ☐

2 Fill in the missing numbers.

a 8, ☐, 10   **b** 3, ☐, 5

c 1, ☐, 3   **d** 5, ☐, 7

e 6, ☐, 8   **f** 7, ☐, 9

3 These numbers have been put in order. Fill in the missing numbers.

a 2, ☐, ☐, ☐, ☐, ☐, 8

b 7, ☐, ☐, 10

4 Order each set of numbers, starting with the smallest.

a 4, 7, 2   ☐, ☐, ☐

b 9, 3, 5   ☐, ☐, ☐

c 6, 10, 8   ☐, ☐, ☐

Date: _____

Lesson 4: **Ordinal numbers**

• Use ordinal numbers to show position

1 Draw lines so the carriages are in order.

| 5th | 9th | 3rd | 6th |

| 1st | 2nd | | 4th | | | 7th | 8th | | 10th |

2 Write the ordinal numbers on the crocodiles, in order.

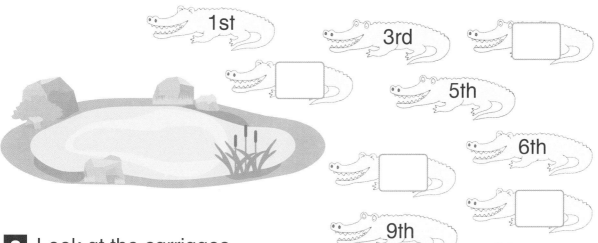

1st

3rd

5th

6th

9th

3 Look at the carriages.
Write the position of each animal.

Date: _____

69

Lesson 1: **Partitioning numbers into tens and ones**

• Partition numbers from 10 to 20 into tens and ones

You will need
• coloured pencils

1 Colour the ten blue and the ones red.

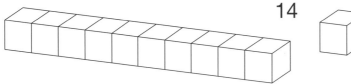

14

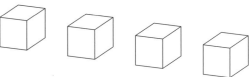

2 Draw tens and ones blocks to match the numbers.

a 15

b 18

c 12

d 13

3 Decompose the numbers into tens and ones.

a

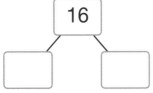

16

b

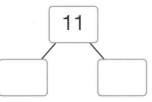

11

c

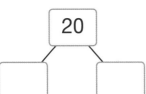

20

d

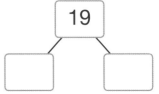

19

Date: _____

Lesson 2: **Combining tens and ones**

- Make numbers from 10 to 20 from tens and ones

1 Write the number shown.

a
10   3

b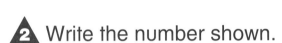
10   8

2 Write the number shown.

a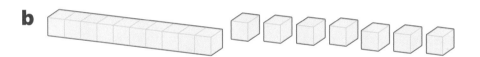

b

c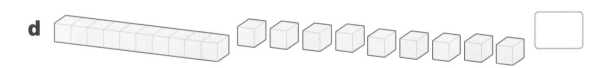

d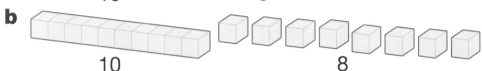

3 Write the number that these numbers make when combined.

a | **10** | **5** | |

b | **10** | **0** | |

Date: _____

Number

Lesson 3: **Representing numbers in different ways**

• Express a number to 20 in different ways

1 Write the matching number sentences.

a  ☐ + ☐ = 17

b  ☐ + ☐ + ☐ = 17

c 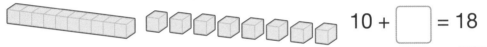 ☐ + ☐ + ☐ = 17

2 How many ways can you think of to regroup 18?

10 + ☐ = 18

10 + ☐ + ☐ = 18   10 + ☐ + ☐ = 18

10 + ☐ + ☐ = 18   10 + ☐ + ☐ = 18

3 How many ways can you think of to regroup 19?

10 + ☐ = 19

10 + ☐ + ☐ = 19   10 + ☐ + ☐ = 19

10 + ☐ + ☐ = 19   10 + ☐ + ☐ = 19

4 How many ways can you think of to regroup 12?
Try regrouping the tens and the ones.

Date: _____

☺ 😐 ☹

Number

Lesson 4: **Comparing and ordering numbers to 20**

• Compare and order numbers to 20

You will need
• counters

1 Use counters to order these numbers from smallest to greatest.

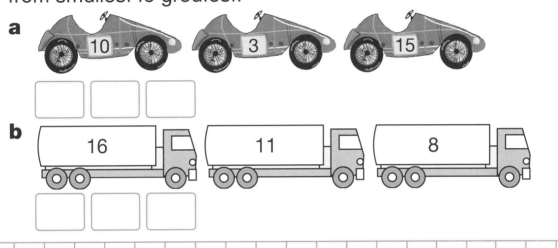

a

b

```
1  2  3  4  5  6  7  8  9  10 11 12 13 14 15 16 17 18 19 20
```

2 Use the number line to order these numbers, smallest to greatest.

a 7 14 12

b 20 3 15

c 17 14 18

d 5 11 10

3 Use the number line to order these numbers, smallest to greatest.

a 6 19 12 4 20

b 8 10 3 16 9

Date: _____

73

Number

Lesson 1: **Halving objects**

- Find half of an object
- Recognise objects that are in halves

1 Draw lines to match the halves.

2 Circle the food that is in halves.

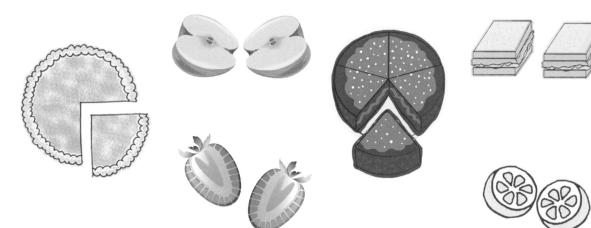

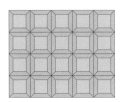 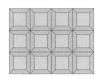 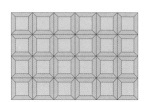

3 Draw lines to halve the chocolate bars.

Date: _____

Lesson 2: **Halves of shapes**

- Find half of a shape
- Recognise shapes that are in halves

1 Tick the shapes that are cut in half.

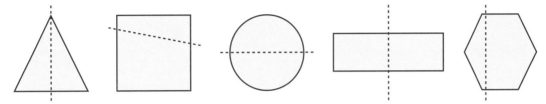

2 Draw lines to match each shape to its half.

3 Draw a line to cut each shape in half.

 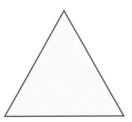  

4 Show four ways you can cut a square in half.

Date: _____

Lesson 3: **Halves of amounts (1)**

Number

• Find half of an amount

1 Circle the groups that have been halved.

2 Share the toys equally. How many does each child get?

a

b

c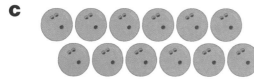

d

3 Find half of each group.

a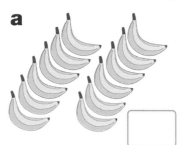

b

c

Date: _____

Lesson 4: **Halves of amounts (2)**

• Find half of an amount

Number

1 Draw spots to make the halves equal.

a 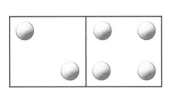 **b**  **c** 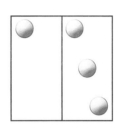 **d**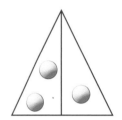

2 Share each group of spots equally. The first one has been done for you.

a 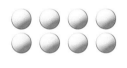 **b** 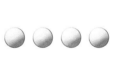 **c**  **d**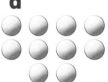

  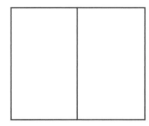

3 Share the cakes equally. How many will each child get?

Date: _____

77

Number

Lesson 1: **Halving numbers to 10**

- Find half of a number to 10
- Use half as an operator

1 Draw spots to find half.

a

b

c

2 Find half of each number.

a Half of 10 ⟶ ☐

b Half of 2 ⟶ ☐

c Half of 6 ⟶ ☐

d Half of 4 ⟶ ☐

e Half of 8 ⟶ ☐

3 Find the starting numbers before they were halved.

a Half of ☐ ⟶ 3

b Half of ☐ ⟶ 5

Date: _____

Lesson 2: **Halving numbers to 20**

- Find half of a number to 20
- Use half as an operator

You will need
- counters or cubes

1 Find half of each number. If you need to, use counters or cubes to help.

a $\frac{1}{2}$ of 4 = ☐   **b** $\frac{1}{2}$ of 2 = ☐   **c** $\frac{1}{2}$ of 6 = ☐

 Find half of each number.

a $\frac{1}{2}$ of 12 = ◯

b $\frac{1}{2}$ of 20 = ◯

c $\frac{1}{2}$ of 16 = ◯

d $\frac{1}{2}$ of 18 = ◯

3 Aliya's age is $\frac{1}{2}$ of Reena's age.

How old is Aliya? ☐   Reena **14**   Aliya

4 Alfie's house number is half of Osman's.
What is Alfie's house number? ☐

18

Osman   Alfie

Date: _____

Number

Lesson 3: **Combining halves (1)**

• Put halves together to make wholes

1 Count the halves.

There are ☐ whole cakes.

2 Count the halves.

There are ☐ whole apples.

3 Count the halves.

There are ☐ whole tomatoes.

4 Count the halves.

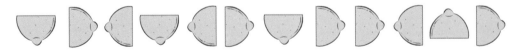

There are ☐ whole lemons.

Date: _____

Lesson 4: **Combining halves (2)**

• Combine halves of an amount back into the whole amount

1 Count each half to find the whole amount.

a $\frac{1}{2}$ 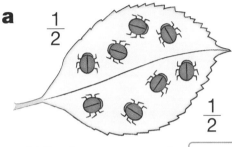 $\frac{1}{2}$

b $\frac{1}{2}$ 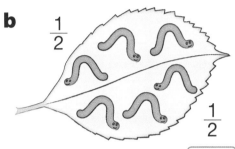 $\frac{1}{2}$

Whole amount is ☐   Whole amount is ☐

2 Half of each amount is shown. Draw the other half.
Count all the dots to find the whole amount.

a ☐

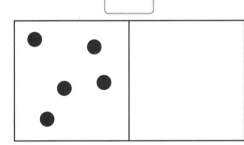

b ☐

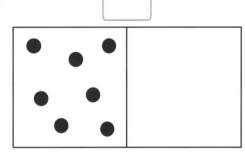

c ☐

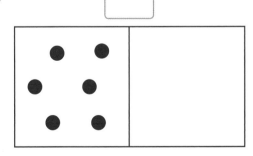

d ☐

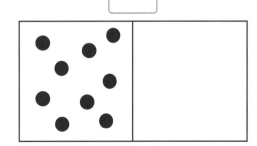

3 If half of an amount is 8, what is the whole amount? ☐

4 If half of an amount is 10, what is the whole amount? ☐

Date: _____

Geometry and Measure

Lesson 1: **Days of the week**

- Know the days of the week
- Understand that there are weekdays and weekend days

1 Draw lines to put the days of the week in order.

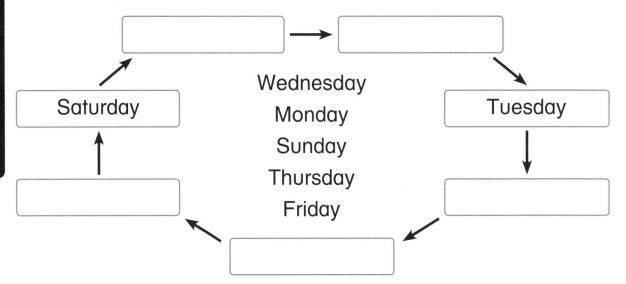

Saturday

Wednesday
Monday
Sunday
Thursday
Friday

Tuesday

2 Write the day that comes next.

a Tuesday, Wednesday,

b Saturday, Sunday,

c Wednesday, Thursday,

3 Write the day before (Yesterday) and the day after (Tomorrow).

Yesterday	Today	Tomorrow
	Monday	
	Friday	
	Wednesday	

Date: _____

Lesson 2: **Months of the year**

- Know the months of the year
- Know some familiar events that happen in each month

1 Draw an event that happens every year. Write the month when it happens.

2 Draw or write five events on this calendar.

January	February	March	April
May	**June**	**July**	**August**
September	**October**	**November**	**December**

3 Write the months in order.

December

February

October

May

Date: _____

83

Lesson 3: **O'clock**

Geometry and Measure

• Recognise o'clock times

1 Tick the clocks that show o'clock times.

2 Write the time each clock shows.

a

_____

b

_____

c

_____

d

_____

3 Draw hands on each clock to show these times.

a 2 o'clock

b 7 o'clock

Date: _____

Lesson 4: **Half past**

• Recognise half past times

1 Tick the clocks that show half past times.

2 Write the time each clock shows.

a

_____

b

_____

c

_____

d

_____

3 Draw hands on each clock to show these times.

a half past 2

b half past 4

Date:

Geometry and Measure

Lesson 1: **Recognising 2D shapes**

- Recognise circles, squares, triangles and rectangles

You will need
- coloured pencils

1 Draw lines to match shapes that are the same.

2 Colour the shapes in the house.

blue triangles

green squares

red circles

purple rectangles

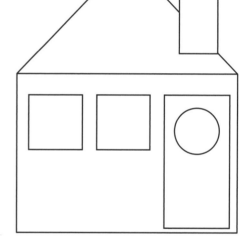

3 How many of each shape?

triangles ☐ squares ☐ circles ☐ rectangles ☐

Date: _____

😊 😐 ☹

Lesson 2: **Describing 2D shapes**

- Recognise the features of circles, squares, triangles and rectangles

You will need
- coloured pencils

1 Trace and count the sides.

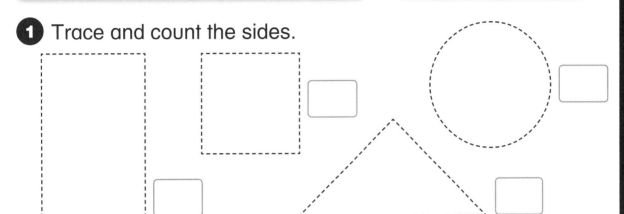

2 Draw lines to join the shapes to the correct box.

| curved side | 4 sides | 3 sides |

3 Colour the shapes that have both curved **and** straight sides.

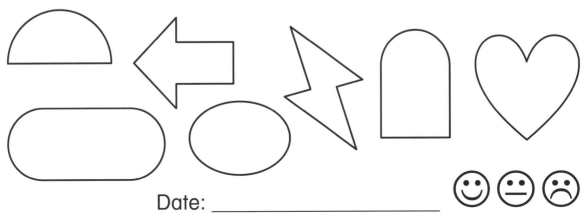

Date: _____   ☺ ☺ ☹   87

Geometry and Measure

Lesson 3: **Sorting 2D shapes**

- Sort circles, squares, triangles and rectangles

You will need
- coloured pencils

1 Colour the shapes that are the same.

a

b

c

2 a Colour all the shapes with curved sides.

b Tick all the shapes with 3 sides.

c Circle all the shapes with 4 sides.

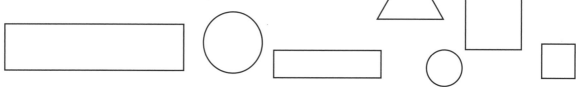

3 List all the ways you can think of to sort these shapes.

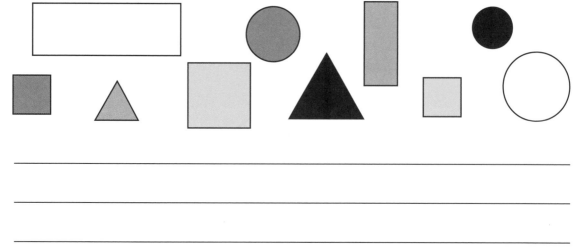

_____

_____

_____

Date: _____

Lesson 4: **Rotating shapes**

- Identify when a rotated shape looks the same
- Describe or extend a repeating pattern

You will need
- coloured pencils
- 2D shapes

1 Do these 2D shapes look the same now that they have been rotated? Circle your answer.

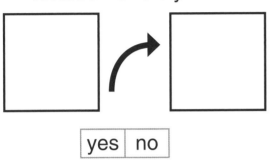 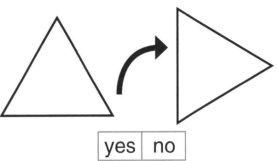

| yes | no |

| yes | no |

2 Colour the shapes that will look the same when they have been rotated. You can use plastic 2D shapes to help you.

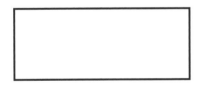 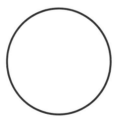  

3 Draw the rectangle in the box to show what it would look like after it has been rotated.

Draw a repeating pattern that uses the rectangle before and after it is rotated.

Date: _____

Lesson 1: **What is a 3D shape?**

- Recognise 3D shapes
- Name 3D shapes

You will need
- coloured pencils

Geometry and Measure

1 Colour the 3D shapes.

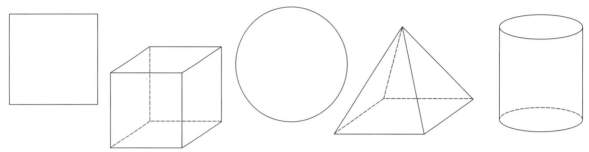

2 Join each 3D shape to its name. Watch out for the 2D shapes!

sphere   cube   cylinder   pyramid   cuboid

3 Circle each 3D shape. Tick each 2D shape.

How many 3D shapes? [ ]   How many 2D shapes? [ ]

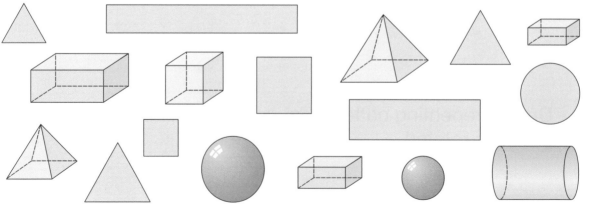

Date: _____

Lesson 2: **Making models with 3D shapes**

- Make models with 3D shapes
- Identify when a rotated shape looks the same

You will need
- coloured pencils

1 Tick the shapes that have been used to make the bridge.

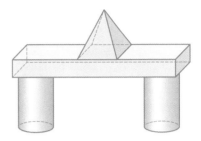

sphere   cuboid   pyramid   cylinder   cube

2 Look at the starting positions of these four shapes.

Starting positions

   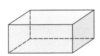

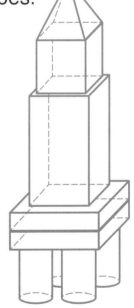

a Colour the shapes in this model that have been rotated.

b Write the names of the shapes that have been used to make this model.

_____

_____

3 Can you tell if the cubes have been rotated in this model? Why/why not?

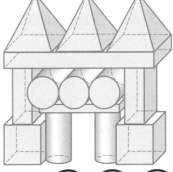

_____

_____

Date: _____

Geometry and Measure

Lesson 3: **Describing 3D shapes**

- Recognise the features of cubes, cuboids, cylinders, spheres and pyramids

You will need
- coloured pencils

1 Colour the shapes with 6 flat faces.

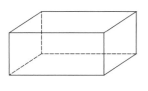 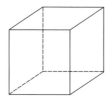 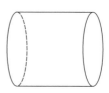 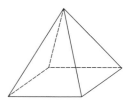

2 Draw lines to match each shape to the number of faces and the description.

3 faces curved

6 faces

1 face curved and flat

5 faces flat

3 Name a shape with 6 flat faces and 12 edges.

_____

4 Name the shape with 2 flat faces and 1 curved face.

_____

Date: _____

Lesson 4: **Sorting shapes**

• Sort 3D shapes

You will need
• coloured pencils

Geometry and Measure

1 Do these shapes roll? ✔ for yes, ✘ for no.

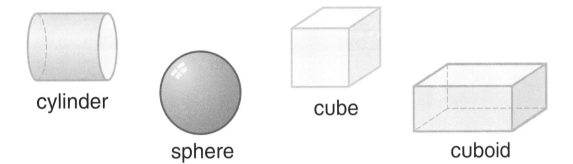

cylinder   sphere   cube   cuboid

2 a Colour all the shapes with curved faces.

b Tick all the shapes with 6 faces.

c Circle all the shapes with 8 or fewer edges.

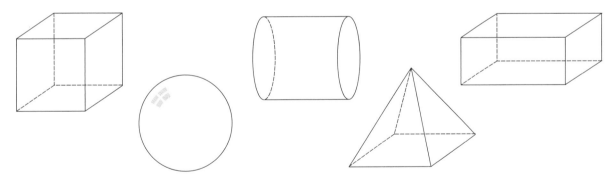

3 How do you think these shapes have been sorted? List two different ways.

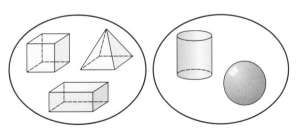

_____

_____

Date: _____

Lesson 1: **Length, height and width**

Geometry and Measure

• Compare and order length, height and width

1 Tick the **longer** caterpillar.

2 Tick the **shorter** caterpillar.

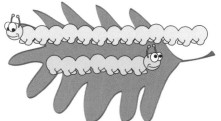
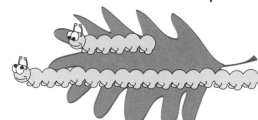

3 Draw a **longer** brush.

4 Draw a **shorter** ribbon

5 Order the animals **shortest** (1) to **tallest** (5).

Date: _____

Lesson 2: **Measuring length**

You will need

- Measure the lengths of objects

- interlocking cubes, glue stick, crayon, pencil, notebook
- uniform non-standing measuring item

1 Use the cubes to measure.

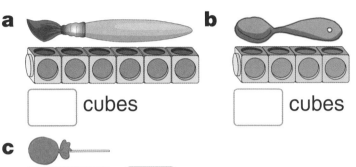

a [ ] cubes

b [ ] cubes

c [ ] cubes

2 Use cubes to measure these things.

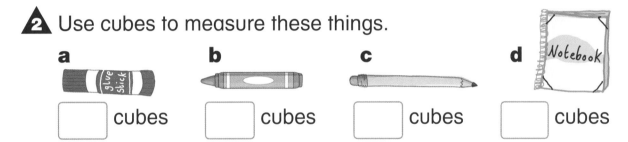

a [ ] cubes

b [ ] cubes

c [ ] cubes

d [ ] cubes

3 Choose something other than cubes to measure these objects with. Then measure each of the objects.

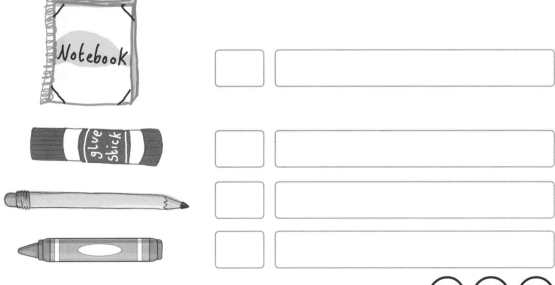

Date: _____

Geometry and Measure

Lesson 3: **Mass**

• Compare the mass of objects

1 Draw a line from each animal to the right word.

heavier

lighter

2 Draw a line from each animal to its correct place on the balance scales.

a

b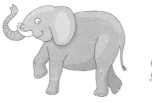

3 Draw something **lighter**.   **4** Draw something **heavier**.

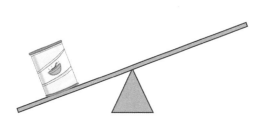

5 Circle the word that is **true** for the square.

a

b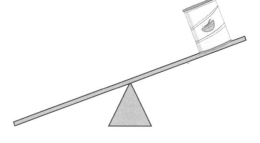

heavier   lighter   same

heavier   lighter   same

Date: _____

Geometry and Measure

Lesson 4: **Measuring mass**

• Use everyday objects to measure mass

1 How many cubes?

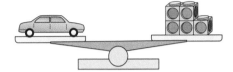 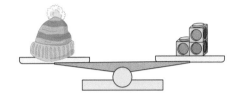

a The toy car has a mass of ☐ cubes.

b The hat has a mass of ☐ cubes.

2 Do you add or take away cubes to make the scale **balance**? Circle your answer.

a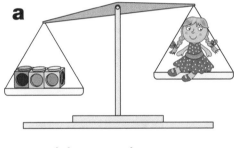

b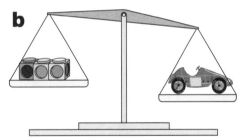

add   take away   add   take away

3 a The 3 bears have a total mass of ☐ cubes.

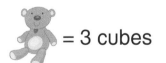 = 3 cubes    = 2 cubes    = 1 cube

b Draw cubes on the scale to make it balance.

Date: _____

Geometry and Measure

Geometry and Measure

Lesson 1: **Full, half full or empty?**

- Recognise when a container is full, half full or empty

You will need
- blue coloured pencil

1 Colour the glass to show it is **full**.

2 Colour to make the labels true.

empty   full   half full

3 Draw lines to the correct labels.

empty   full   half full

Date: _____

Lesson 2: **Estimating and comparing capacity**

Geometry and Measure

• Estimate and compare capacity

1 Circle the container that holds the **most**.

2 Order the capacities from **least** (1) to **most** (4).

a

b

3 Draw the containers from **2 a** in order, starting with the container that holds the **most**.

Date: _____

Geometry and Measure

Lesson 3: **Measuring capacity**

You will need
- yoghurt pot
- jug
- small bucket
- water

- Measure capacity in units of measurement that are the same

1 Use the yoghurt pot to fill the jug with water.

The capacity of the jug is [ ] yoghurt pots.

2 Use the jug to fill the bucket with water. How many jugs did you need to fill the bucket?

Draw them.

3 Which is the best container to use to fill the paddling pool? Circle it.

Date: _____

Lesson 4: **Temperature**

- Identify when we feel hot or cold
- Recognise when an object is hot or cold

You will need
- red and blue coloured pencils

1 Ted is going on holiday. Will he feel **hot** or **cold**?

a

b

c

     hot   cold   hot   cold   hot   cold

2 Colour the **hot** things red and the **cold** things blue.

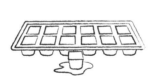

3 Order these things from **coldest** (1) to **hottest** (4)

Date: _____

101

Geometry and Measure

Lesson 1: **Describing direction**

• Give directions to move an object

1 Draw lines to match the picture and the word.

up

down

2 Where is Ted going? Circle the correct word.

forwards   backwards

forwards   around   up

up   down

3 The snail moves **forwards** to the flower. It then moves **around** the flower. Then it moves **up** the hill, **down** the hill and **forwards** to the apple. Draw a line to show the snail's journey.

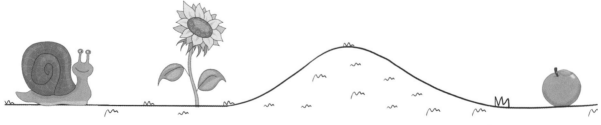

Date: _____

Lesson 2: **Left and right**

• Know which way is left and which way is right

 Tick the **left** hand.

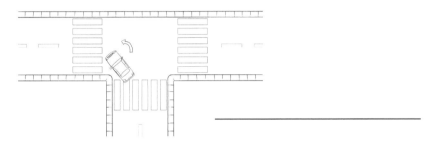

2 a Which way is the car turning?

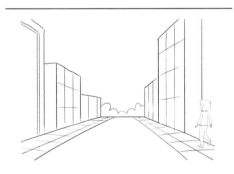

_____

b Which side of the road is the man on?

_____

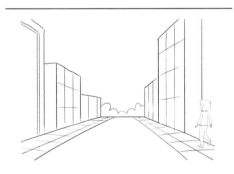

c Hassan puts a cross (✗) in the space to the **left** of the circle. Draw his cross.

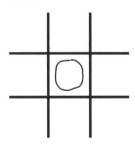

3 Put your pencil on the dot. Now follow the directions:

• Move your pencil up.

• Move it right.

• Move it down.

• Move it left.

• Draw a ▲ where your line ends.

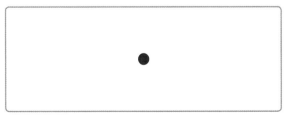

Date: _____

Geometry and Measure

Lesson 3: **Describing position**

• Describe an object's position

1 Draw

 in

 on

2 Where is the cat?

on

outside

behind

under

inside

3 Draw

 over

 under

 inside

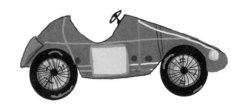

Date: _____

Geometry and Measure

Lesson 4: **Following directions**

• Follow directions

1 Draw a line to show the snail's movements:

- **right**
- **up** the tree
- **down** the tree.

2 Draw a line to show the bird's movements:

- **up** the **left** side of the post
- **around** the light
- **down** the **right** side of the post.

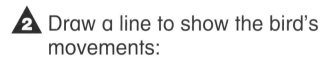

3 The ladybird followed these directions but got one of them wrong. Circle the direction it got wrong.

- Go forwards
- Go left
- Go forwards
- Go right
- Go forwards
- Go right

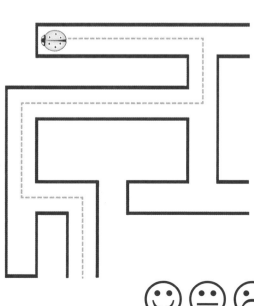

Date: _____

Statistics and Probability

Lesson 1: **Sorting data**

• Sort data

1 Draw lines to sort the buttons.

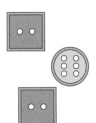

round

square

2 Are there more round or square buttons? Tick the answer.

round	square

3 Draw lines to sort the buttons.

2 holes

6 holes

4 Are there more buttons with 2 or 6 holes? Tick the answer.

2 holes	6 holes	the same

5 Write the letters and numbers in the correct box.

Letters	Numbers

A   3   B   J   5   10   T   9   K   P   6   N   W   4   L   7

6 Are there more numbers or letters? Tick the answer.

numbers	letters

Date: _____

Lesson 2: **Collecting data**

> • Collect data to answer a question

1 Favourite animals

   a How many elephants? ☐

   b How many lions? ☐

   c How many tigers? ☐

2 Favourite foods

   a How many apples? ☐

   b How many cakes? ☐

   c How many ice creams? ☐

   d How many bananas? ☐

   e What is the most popular food?

_____

3 Omar wants to know how children in his class travel to school. He knows that some walk. He knows that some go by car. He knows that others cycle. Think of two ways that Omar could find out.

_____

_____

_____

Date: _____

Statistics and Probability

Lesson 3: **Using lists**

• Organise data in a list

1 Complete the list.

Animals I saw in the zoo

2 a Complete the list.

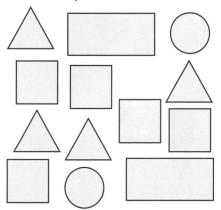

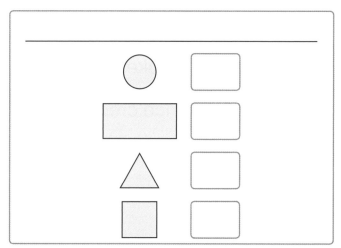

b What was the most popular shape? _____

3 a What question do you think this list is answering? _____

_____

b What would be a good title for this list? _____

_____

Date: _____   ☺ 😐 ☹

Lesson 4: **Using tables**

• Organise data in a table

1 Complete the table.

Sweet	Total

2 Complete the table.

Vehicle	Total

3 Complete the table.

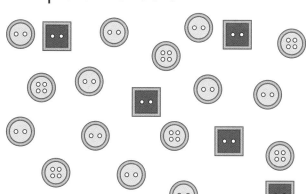

Button type	Total

Date: _____

Statistics and Probability

Statistics and Probability

Lesson 1: **Venn diagrams**

• Use a Venn diagram to sort information

1 Write the numbers in the diagram.

1

3

4

7

10

Numbers

even
numbers

2 Complete the Venn diagram.

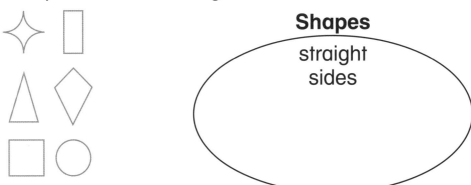

Shapes

straight
sides

3 Complete the label.

Animals

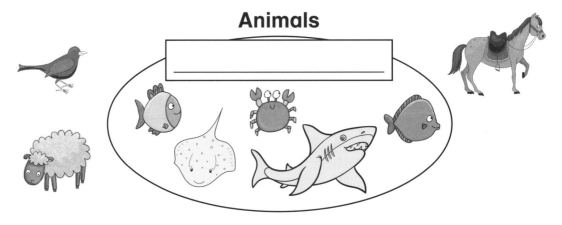

Date: _____

Lesson 2: **Carroll diagrams**

• Use a Carroll diagram to sort information

1 Complete the Carroll diagram.

A B C
D E

Letters with curves	Letters with no curves

2 Complete the Carroll diagram, then answer the questions.

8 H K 12 9 T
4 16 C 20 P

Letter	Not a letter

How many are letters? ☐

How many are not letters? ☐

3 Complete the Carroll diagrams.

Less than 10	More than 10

Even number	Odd number

Date: _____

Statistics and Probability

Lesson 3: **Pictograms**

• Use a pictogram to show information

You will need

• coloured
  pencils

1 Write the totals.

Cakes sold		Total
Monday	🧁🧁🧁🧁🧁🧁🧁🧁🧁	
Tuesday	🧁🧁🧁🧁	
Wednesday	🧁🧁	

2 Write the totals.

Safari park		Total
giraffe	(4 giraffes)	
zebra	(3 zebras)	
tiger	(6 tigers)	
wolf	(8 wolves)	
bear	(2 bears)	

3 Use the table to complete the pictogram.

Favourite fruit

banana	
apple	
orange	

Fruit	Number
banana	2
apple	5
orange	3

Date: _____

Lesson 4: **Block graphs**

You will need
- coloured pencils

- Use a block graph to show information

1 Complete the block graph.

Fish in the pond

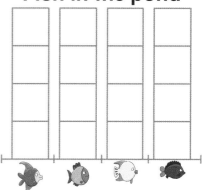

2 Complete the block graph. Then write a statement about the block graph.

Pencils in the pot

_____

_____

_____

red
8

blue
10

green
5

red   blue green

Our favourite vegetables

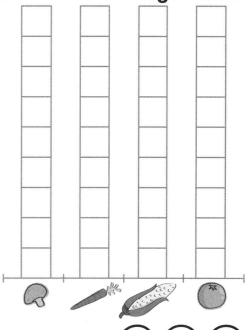

3 Use the table to complete the block graph. Use blue to colour the column for the most popular vegetable.

🍅	7	🥕	9
🍄	5	🌽	6

Date: _____

😊 😐 ☹️

Statistics and Probability

Acknowledgements

Photo acknowledgements

Every effort has been made to trace copyright holders. Any omission will be rectified at the first opportunity.

p15l Kristyna Vagnerova/Shutterstock; p15c Vtaurus/ Shutterstock; p15r Lenin Graphics/Shutterstock; p23tl Rvector/ Shutterstock; p38t Diana Riabets/Shutterstock; p38ct Robuart/ Shutterstock; p38cb Diana Riabets/Shutterstock; p38b Okili77/ Shutterstock; p62cl Brovko Serhii/Shutterstock; p62cr Peyken/ Shutterstock; p62tr Tuulijumala/Shutterstock; p62l AVIcon/ Shutterstock; p62b Miniyama/Shutterstock; p69t Mr. Luck/ Shutterstock; p84t Szefei/Shutterstock; p84b Rangizzz/ Shutterstock; p85t Szefei/Shutterstock; p85b Rangizzz/ Shutterstock; p98t Onyxdesigns.Std/Shutterstock; p102t BlueRingMedia/Shutterstock; p108b Rwdd_studios/ Shutterstock; p110bl Kristina07/Shutterstock; p110br Ksenya Savva/Shutterstock.